CHRIS FERRIE

THE CAT IN THE BOX

ILLUSTRATED BY KEVIN SHERRY

sourcebooks

To the future quantum physicist who will disentangle our quantum mess. –CF

To my mom, Cathy Sherry, who taught kids to love math. –KS

Text © 2019 by Chris Ferrie
Illustrations © 2019 by Kevin Sherry
Cover and internal design © 2019 by Sourcebooks, Inc

Art was crosshatched with fine-tipped felt markers during the brief moments when my cat would stop snoozing all over my bristol board.

Published by Sourcebooks Inc.,
P.O. Box 4410, Naperville, Illinois 60567-4410
(630) 961-3900
Fax: (630) 961-2168
sourcebooks.com

Library of Congress Cataloging-in-Publication Data is on file with the publisher.

Printed and bound in China.
LEO 10 9 8 7 6 5 4 3 2 1

It did not sit well.

It was **far** too confusing.

So we sat at the blackboard,

the two of us musing.

I sat with Schrödinger,

dreaming of quantum theory,

but the **math** of it all

made both of us dreary.

What was it about?

We were not content.

The trouble came down to

the **entanglement**.

And then,

a knock at the door.

A **box**,

but who was it for?

We looked in!

And saw a **cat** in a **box**!

We looked in!

And saw a **cat** that **talks**!

And it said to us,

"Get me out of this box!"

I was happy to see it—

an extra **mind**.

Right now we needed

all the **help** we could find.

And then, an idea:

"With this cat's great sight,

it could see a single quantum.

A quantum of light!

An excited **atom** will decay—

a random condition.

The cat might see

a **superposition**!"

Schrödinger went on,
"Next we close the lid.
To **look** in the box
is what I forbid.
The cat will see…

and not see.
An odd condition.
The cat will be
in **superposition**!"

Schrödinger used
this cat in a box
to dream up the first
quantum paradox.
A paradox is something
that doesn't make sense.
There must be an assumption
that is causing offense.

"The theory has failed!"
Schrödinger cried.
"We must start all over
and swallow our pride."

But then I smiled
to our own merriment.
We have the cat, box, and atom
to do the **experiment**.
The scientific method
is to test hypotheses,
so our knowledge can grow
beyond just the ABCs.

Through all of this,
shaking its head,
the cat quietly listened
to all that was said.

"A quantum superposition,"

said the cat with a sigh.

"A silly idea!

and I'll tell you why.

But I must teach you some **physics**

and where it applies!"

The lesson began, "I see your intent.

But the most important thing

is the **measurement**.

When I look at the atom

to see if it's spent,

I create in the world

a **quantum** event.

When measurement acts,

the atom itself

is forced to collapse.

It might stay **excited**.

It might **decay**.

But which it will be,

no one can say."

"But all is not lost.

We are not throwing dice.

We can calculate **some** things

when science gives us advice.

Though nothing is certain,

one thing is key.

Physics can tell us

the **probability**.

Quantum probability—

a tricky **concept**.

To use it, one needs

to be quite adept.

But study the math,

and before long,

the quantum tune

you'll be singing along.

Algebra, **calculus**, and **geometry**.

The more math you know,

the happier you'll be."

$$F^1 \; 26/\sqrt{}$$
$$B) \; 34 \lor \boxed{35°} \; L$$
$$+\frac{A}{C}) \; Z = N^2(A,B)(C,$$
$$P) \; GM, M2 \; A^2 + B^2 =$$
$$120°)F = \frac{}{D}$$
$$C(S-1)^F + rS'nO(o)^2M$$

The cat was not done.

There was more quantum physics.

Such a rich theory

can't be taught in mere minutes.

"The next lesson I have

is where the name **comes** from.

I will tell you why

we call it **quantum**."

"You see," the cat said,

"when you measure my size,

any number at all

is due to arise."

"But measure an atom

and what will come out?

A single quantum of energy

will be brought about.

A **photon** of light

will be let go.

The atom's energy

from high to low.

The photon's energy

must be exact.

One quantum of light

from the atom subtract."

"So you see now,"
said the boxed cat.
"No superposition.
No doubt about that.
When I measure, I **break**
the quantum effect.
No paradox at all
that you should suspect."

"I know all of this,"
Schrödinger quickly replied.
"I **invented** this theory,
a fact none deny.
I'm afraid it is *you*
who is missing a fact.
The best part of the theory
is due for its act.
Of course I am talking
of entanglement—
the source that is giving me
such bafflement."

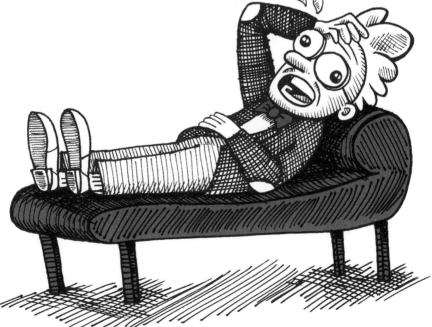

"The box will be closed.
Just the atom and you.
A large quantum thing
becomes of the two.
You and the atom—
entanglement!

A **curious** state.

It's no accident.

We'll open the box

and ask just one question.

What did it look like—

the quantum connection?

I wish to resolve

this deep mystery.

So **future** scientists

have less misery."

The cat disagreed.

"It's not **mysterious**.

You simply have mixed up

our experience.

Outside the box

lies **uncertainty**.

But for me in the box,

I will know what I see.

The light from the atom

will come or will not,

and not **superposed**

as you have thought."

But the cat in the box

could see without doubt

we were not ready

to let it get **out**.

So the cat agreed

to stay in and wait

for the atom inside

to change its state.

So the trial began,

the atom and cat.

Inside the box,

it quietly sat.

We waited too.

The tension was rife.

We decided to wait

for the atom's **half-life**.

That is the time

for the atom's decay

to reach the point

of half what it may.

The point of half-life

was shown on the clock.

Schrödinger knew

what went on in the box.

Two **possibilities**

at the same time—

but open the box

and only one we shall find.

But what had gone on
before the box opened?
From the cat in the box
it will be spoken.
So we opened the box,
and what was in there?

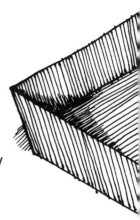

The cat sitting quietly
licking its hair.

"**Cat!**" shouted Schrödinger.

"What do you say?

What did you see?

Do not delay."

"It's as I said.
I saw the **photon**.
Nothing strange at all
in the box has gone on.
The atom **collapsed**
a short time ago.
I measured its light
as it will now show."

For science to work,

repetition is key.

To correct errors

we need at least three.

But more is better.

Data is gold.

So **valuable** that

it's bought and it's sold.

So the cat went back in
again and **again**.

The cat was amused
and did not complain.

The results were in
and we had to agree.
There was no **quantum magic**
for the cat to see.
The **light** from the atom
came or did not.
But never them both
as Schrödinger thought.

Now we are **happy**

with no paradox,

thanks to the cat…

the cat in the box.

AUTHOR'S NOTE

Erwin Schrödinger was an Austrian (not Australian!) physicist who made seminal contributions to the development of quantum theory. Within physics, he is most famous for his wave equation, called the Schrödinger Equation, which accurately describes how quantum things change in time. Outside of physics, he is known for his thought experiment, affectionately known as Schrödinger's Cat.

But don't get too comfortable—the original idea was to have the hypothetical cat sit in a box with a vial of poison! Next to the poison was a hammer. The hammer was held above the vial so delicately that one single atom decaying would drop it. Unfortunately for the cat, there was also an excited atom in the box. Just as our cat would "see and not see," Schrödinger said his cat would be "dead AND alive." Quite paradoxical!

Quantum theory describes a world much different from the one we experience. The prevailing view at the time was to not worry about it because the theory only applied to things so small (atoms, etc.) that their counterintuitive effects would never enter our everyday experience. Schrödinger thought otherwise. To demonstrate why it was important to worry about what quantum theory implies for our experience, he invented the cat. Happily for cats, the experiment was never performed! Along the way, Schrödinger also coined the term "entanglement," which we now understand to be one of the most difficult aspects of quantum theory to grasp.

The most common resolution of the paradox is to treat the cat as another scientist who privately uses quantum theory to describe the physics going on in the box. In doing so, the cat will make quantum measurements and will accurately predict the outcomes of experiments. The cat doesn't really care what Schrödinger is doing outside the box and all the trouble he is having. Quantum theory works perfectly well for everyone describing the observations they themselves make.

For some, this is still unsatisfying and only introduces more questions! What counts as a scientist? Can a cat without agency be something or someone that "uses" quantum theory? Will we ever be comfortable in our understanding of entanglement as it applies to everyday objects? These are all questions that scientists are trying to answer. Perhaps you have some of your own, or maybe more questions. No matter—everyone is welcome!

CHRIS FERRIE is an award-winning physicist and researcher at the University of Technology Sydney and Centre for Quantum Software and Information. He obtained his PhD in applied mathematics from the Institute for Quantum Computing, University of Waterloo (Canada). Occasionally, he writes children's books. Chris lives in Australia with his wife and children.

KEVIN SHERRY is the author/illustrator of the The Yeti Files series and the award-winning picture book, *I'm the Biggest Thing in the Ocean*. Kevin is also a puppeteer, a screen printer, and a zine creator. He was ultimately discovered at a small press expo. Kevin lives and works in Baltimore with his cat, Kinkos.

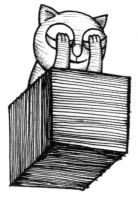